Everyday Health

NUTRITION

GLOBE FEARON
Pearson Learning Group

Reviewers

Patricia Williams, Ed.D., R.N.
Mount Vernon Public Schools, Mount Vernon NY 10553

Sheila Nigh
Wichita Public Schools, Wichita KS 67203

Lois Lanyard
Woodbridge High School, Woodbridge NJ 07095

Linda Jacobs, Ed.D.
The Harbour School, Annapolis MD 21401

Barbara Young
Bryan Station High School, Lexington KY 40505

Credits

Project Editors	Douglas Falk, Laurie Golson, Stephanie Cahill, Jennifer McCarthy
Executive Editor	Joan Carrafiello
Market Manager	Margaret Rakus
Assistant Market Manager	Donna Frasco
Editorial Development	WordWise, Inc.
Production Editor	John Roberts
Production Director	Kurt Scherwatzky
Interior Design	Pat Smythe
Illustrators	Pat Carroll, Daisy de Puthod
Electronic Page Production	Lesiak/Crampton Design
Cover Design	Pat Smythe
Cover Art	James McDaniel
Editorial Assistant	Eugene Myers

Photo Credits: p. 4: © Glyn Cloyd; p. 18: © Glyn Cloyd; p. 34: Steven Whalen, Zephyr Pictures.

Copyright © 1997 by Pearson Education, Inc., publishing as Globe Fearon®, an imprint of Pearson Learning Group, 299 Jefferson Road, Parsippany, NJ 07054. All rights reserved. No part of this book may be reproduced or transmitted in any form or by any means, electronic or mechanical, including photocopying, recording, or by any information storage and retrieval system, without permission in writing from the publisher. For information regarding permission(s), write to Rights and Permissions Department.

ISBN 0-8359-3371-7
Printed in the United States of America
6 7 8 9 10 11   07 06 05 04 03

1-800-321-3106
www.pearsonlearning.com

TABLE OF CONTENTS

UNIT ONE	BASIC NUTRITION		2
Lesson 1	Understanding Nutrients		2
Lesson 2	Vitamins, Minerals, and Water		8
Lesson 3	Calories and food Energy		14
Unit Review			18
UNIT TWO	MANAGING YOUR MEALS		20
Lesson 4	Choosing a Balanced Diet		20
Lesson 5	Choosing the Right Food		26
Lesson 6	Meal Planning		30
Unit Review			36
UNIT THREE	MANAGING YOUR WEIGHT		38
Lesson 7	Weight and Nutrition		38
Lesson 8	Losing Weight Safely		44
Lesson 9	Avoiding Fad Diets		50
Unit Review			54
GLOSSARY			56
RESOURCES			58

TO THE STUDENT

Welcome to *Everyday Health!* By using this series of books, you will achieve these goals:

- You will be able to set goals and make decisions about your health.
- You will be able to help your family and your community in health matters.
- You will be able to practice behavior to improve health.
- You will be able to reduce risks to your health.
- You will understand your role in preventing disease.
- You will understand how to use health information about products.
- You will be able to communicate with others about improving health.

EVERYDAY HEALTH

NUTRITION

This book will help you understand what nutrition is all about.

You probably know all the foods shown above. Perhaps you are planning to eat some of them today.

Do you know the importance of these foods to your nutrition? Which of these foods are good sources of protein, vitamins, and minerals? Can you plan a balanced meal using the foods shown above? Lessons in this book have the information to answer these questions.

When you have finished this book, you will be able to plan a balanced diet, and make decisions about how much food you should eat. You'll know about safe ways to lose weight, and how many calories you need each day. You'll even be able to use the Nutrition Facts labels on food packages to plan healthy snacks.

UNIT ONE • BASIC NUTRITION

Lesson 1

Understanding Nutrients

A healthy afternoon snack can include juice, fruit, crackers, or cheese.

Lesson Objectives

You will be able to

- explain the importance of carbohydrates, proteins, and fats in maintaining the body.
- identify foods that contain carbohydrates, proteins, and fats.

Words to Know

amino acids	units that combine to form proteins
carbohydrates	nutrients that are the body's main source of energy
cells	the smallest parts that make up a living thing
fats	nutrients that supply the body with concentrated energy
nutrients	the substances in food that your body uses to grow and work properly
nutrition	all the ways the body takes in and uses nutrients to stay healthy
proteins	nutrients that help build and repair cells and tissues in the body

2 Lesson 1

Kristin, Olivia, John, and Wayne were hungry. They had stayed late after school to help decorate the gym for the big dance. They hadn't eaten anything since lunch—and that was more than four hours ago.

"It's too bad we don't have any money," said Kristin. "I could really go for a large pepperoni pizza right now."

"I know what you mean," replied Wayne. "I haven't had pizza for ages. My mom and dad are on some kind of health kick. All we have to eat around here is fruit and vegetables."

"That's all my sister eats all the time," said John.

"Stop complaining," said Olivia. "I'm so hungry even these grapes taste great. And we all know this food is better for us than soda and potato chips."

Your Need for Food

You eat food every day. You eat when you are hungry. You eat for enjoyment. You also eat to spend time with your family and friends. In fact, you will spend about one tenth of your life eating.

Why do you really eat food? You eat food because your body needs energy to do the things you do. Your body also needs food to work properly. Food provides the energy your body needs. Food also contains **nutrients**. Nutrients are the substances in food that make your body grow and work properly.

Your body needs six kinds of nutrients. You will learn about three of these nutrients in this lesson. These nutrients provide energy and substances for your body. They are **carbohydrates**, **fats**, and **proteins**. Three other kinds of nutrients help your body work properly. They are vitamins, minerals, and water. You will learn about these three kinds of nutrients in Lesson 2.

To be sure your body stays healthy, you need to know about how it takes in and uses nutrients. **Nutrition** is the study of nutrients in food and how your body uses them.

Nutrition in Action

1. Food provides your body with _____ and _____.

2. Nutrients that provide energy and substances for your body include _____ , _____ , and _____ .

3. Why should you study nutrition?
   _____
   _____

Nutrients give your body energy

Two different kinds of nutrients give your body energy. Carbohydrates are nutrients that are the body's main source of energy.

There are two kinds of carbohydrates. *Sugars* are simple carbohydrates that your body can use with little or no change. Fruits, fruit drinks, cookies, jelly, and honey all contain sugars. *Starches* are complex carbohydrates. A starch is made of many simple sugars linked together in a long chain. Rice, potatoes, noodles, bread, and dried beans contain complex carbohydrates.

Simple carbohydrates can be quickly used by your body. Complex carbohydrates must be broken down by the body to be used for energy. Some foods that are high in sugars, such as candy or jelly, contain few nutrients your body needs. Foods that contain complex carbohydrates, such as bread or beans, contain vitamins or other nutrients your body needs.

Fats provide the body with concentrated energy. Fats provide about twice as much energy as the same amount of carbohydrates. Meats, poultry, fish, nuts, cheese, cream, whole milk, butter, margarine, mayonnaise, vegetable oil, and olive oil contain fats. Fried foods also contain fats because they are cooked in fat.

There are two kinds of fats. One kind comes from animals. The other kind comes from plants. Fats from animals include butter, whole milk, cheese, egg yolks, and shellfish. Many of these fats are hard at room temperature. Eating too many of these fats can cause serious diseases of the heart and blood vessels. Fats from plants include oils such as peanut oil, corn oil, sunflower oil, safflower oil, and olive oil. These fats are usually liquid at room temperature. You should try to replace animal fats with plant fats as much as possible in the foods you eat.

Complex carbohydrates come from foods such as bread, pasta, grain, and potatoes.

Dairy foods are sources of animal fats.

Corn oil, vegetable oil, and nuts are sources of plant fats.

Sometimes you cannot use all the fats you eat. You store the unused fat in your body. Some stored fat is good for you. It protects your internal organs from damage. Some fat stored under your skin helps keep you warm. But too much stored fat in the body makes you overweight. Being overweight makes your body work harder and is not good for your health. Most people need to eat only a small amount of fat each day.

Think About It

4. What is the difference between a simple carbohydrate and a complex carbohydrate?

_____
_____
_____
_____

5. In what two ways is stored fat good for your body?

_____
_____
_____

Nutrients help you grow

Your body needs nutrients to help it grow and remain strong. Your body is made of tiny units called **cells**. Proteins are nutrients that help build and repair cells in the body. Proteins help build muscles and bones. Proteins are also used to build brain and blood cells. In fact, most cells in your body contain many kinds of proteins.

Proteins are found in many different kinds of food. Meats, chicken, and fish contain protein. Eggs, milk, cheese, dried beans and peas, and nuts also contain protein.

Proteins are made up of small units called **amino acids**. Your body needs 22 different amino acids. Your body makes most of the amino acids it needs. But it cannot make 8 amino acids. They have to come from the foods you eat. All 8 of these amino acids are found in meat and foods that come from animals.

Some people follow a vegetarian diet. They eat only plants. They do not eat any meat or meat products. Plants contain many amino acids, but no single plant contains all of them like meat. People who eat vegetarian diets must combine different kinds of plant foods, including beans and rice, to get all the amino acids they need.

Lean meat, fish, beans, milk, and eggs are sources of protein.

Lesson Review

Vocabulary Review

Write the number of the definition from Column 1 on the line in front of the correct term in Column 2.

Column 1
6. the substances in food that your body uses to grow and work properly
7. nutrients that help build and repair cells and tissues in the body
8. all the ways the body takes in and uses nutrients to stay healthy
9. nutrients that are the body's main source of energy
10. nutrients that supply the body with concentrated energy

Column 2
___ a. carbohydrates
___ b. fats
___ c. nutrients
___ d. nutrition
___ e. proteins

6 Lesson 1

Fill in the blanks in the following statements with the correct word or phrase.

11. An example of a simple carbohydrate is _____. An example of a complex carbohydrate is _____.

12. The six basic nutrients your body needs are _____, _____, _____, _____, _____, and _____.

13. Animal fats are usually _____ at room temperature. Plant fats are usually _____ at room temperature.

Answer the following questions on the lines provided.

14. How are carbohydrates and fats similar? How are they different?

_____
_____
_____

15. Why is it important for a person who eats a vegetarian diet to combine foods such as beans and rice?

_____
_____
_____

Portfolio

What Would You Do?

16. Work with a partner to develop a skit. In your skit, imagine that one of you eats a vegetarian diet. You like to eat only fruits and vegetables. You do not like to eat breads, noodles, or beans. Explain to your friend why it is important to combine different kinds of foods such as rice and beans in the diet.

Lesson 2

Vitamins, Minerals, and Water

You usually do not need to take vitamin pills. You can get all of the vitamins you need by eating a variety of healthy foods.

Lesson Objectives

You will be able to

- explain the importance of vitamins, minerals, and water in maintaining the body.
- identify foods that contain vitamins, minerals, and water.

Words to Know

vitamins nutrients that help the body use other nutrients

minerals nutrients that help the body grow and function normally

oxygen a gas in the air you breathe

Janice and Karen were confused. They read in a magazine that they needed **vitamins** to stay healthy.

"Which one of these vitamins is the best?" asked Janice. "There are so many to choose from."

"I don't really know," answered Karen. "Some of them have just one vitamin. Others have lots of vitamins in one jar. Are we supposed to have minerals, too?"

"I think so," said Janice. "Before we buy anything, we better talk to someone who knows something about vitamins and minerals."

Vitamins Help Your Body Work

Your body needs **vitamins** each day. But, you usually do not need to take vitamin pills. You can get most of the vitamins you need from the foods you eat.

Your body needs only small amounts of vitamins. Vitamins are nutrients that help your body use other nutrients. Vitamins help your body use carbohydrates, proteins, and fats. Vitamins are important for many things. They help you see better and have healthy skin and hair. Vitamins also help your muscles and nerves work better. The table below shows some of the important vitamins, how your body uses them, and in what foods they are found.

Lack of vitamins

A lack of vitamins in your diet can cause health problems. A lack of vitamin A can cause you to have difficulty seeing in dim light. A lack of vitamin B_2 can interfere with proper growth and cause eye, skin, and blood problems. A lack of vitamin C causes teeth to loosen and gums to bleed. A lack of vitamin D causes problems in bone formation and results in bowed legs.

Vitamin	Uses in the Body	Sources
A	Keeps skin and hair healthy, helps you see better at night, keeps bones strong	Green and yellow vegetables, carrots, milk, liver
B_1 (thiamin)	Keeps nervous system healthy, helps body get energy from carbohydrates	Pork, liver, poultry, whole grains, cereals and breads, peanuts
B_2	Keeps skin healthy, helps body get energy from food	Milk, eggs, liver, green leafy vegetables, fish, lean red meat
B_3 (niacin)	Keeps skin, nervous system, and other tissues healthy; helps body get energy from food	Fish, beans, peas, eggs, meat
C	Keeps gums, muscles, blood vessels, and skin healthy, helps heal wounds, helps build strong joints between bones	Citrus fruits, tomatoes, dark green vegetables, berries
D	Helps body use calcium, keeps bones and teeth strong	Milk, eggs, liver; body makes vitamin D in sunshine

Nutrition in Action

1. What vitamin(s) help keep your gums and skin healthy?

2. What foods are good sources of vitamin A?

3. How is vitamin B_3 used in the body?

Minerals Make Your Body Strong

Your body also needs **minerals**. Minerals are nutrients that help make your body grow and work properly. Minerals help build strong bones and teeth. Minerals also keep nerves and muscles working properly.

Four of the most important minerals your body needs are calcium, iron, iodine, and phosphorus. Your body has more calcium than any other mineral. Iron is a mineral in your blood. It helps carry **oxygen** to all your body's cells. Oxygen helps your cells use the energy in food.

Your body also needs many other minerals such as copper, potassium, sulfur, and sodium. Your body needs these minerals in small amounts. Like vitamins, most of the minerals your body needs come from the foods you eat. The table on the following page shows some (but not all) important minerals, how your body uses them, and in what foods they are found.

Lack of minerals

A lack of minerals in your diet can cause health problems. A lack of calcium can make bones weak and more likely to break. A lack of iodine can cause a gland in your throat to become larger. A lack of iron can cause a person to feel weak, tired, and dizzy. Some doctors think a lack of copper is linked to heart disease.

Mineral	Uses in the Body	Sources
Calcium	Helps make strong bones and teeth, helps muscles and nerves work	Milk and dairy products, tofu, soybeans, dark green leafy vegetables
Iodine	Helps control the body's use of energy	Seafood, iodized salt
Iron	Helps blood carry oxygen	Liver, egg yolks, green leafy vegetables, dried fruits, whole-grain breads and cereals
Phosphorus	Helps make strong bones and teeth, helps cells work properly	Meat, poultry, fish, eggs, nuts, whole-grain cereals, milk, broccoli
Potassium	Helps nerves and muscles work	Baked potatoes, bananas, oranges, lima beans, peanut butter
Zinc	Helps growth, keeps brain alert	Lean red meat, pork, liver, eggs, fish, oysters

Think About It

Answer the following questions on the lines provided.

4. How are minerals important to your health?

5. Why do you need milk and dairy products in your diet?

6. What problem would result if you did not have enough iron in your diet?

7. What foods should you eat to make sure your muscles and nerves work properly? Why?

Water is a very important nutrient. More than six-tenths of your body weight is water. You need to drink water every day.

Water Is an Important Nutrient

Water is a very important nutrient. It is essential for staying alive. You can survive for about two weeks without eating food. But, you can survive for only two or three days without water.

Every part of your body contains water. All the cells in your body contain water. Your blood is about 80 percent water. Even your hard bones contain about 30 percent water. About 65 percent of the total weight of your body is water.

Your body uses water in many ways. Water helps break down food and carry away wastes. Sweat, which is mostly water, helps control your body's temperature.

Your body loses more than 2 quarts of water each day. You lose water when you sweat. You also lose water each time you breathe out, or exhale. Your body also loses water in body wastes.

You need to replace the 2 quarts of water you lose each day. You can do this by drinking eight 8-ounce glasses of water each day. Soup, milk, fruit juice, tea, vegetables, and fruits all contain water. You can tell when your body needs water. You feel thirsty.

Lesson Review

Vocabulary Review

On the lines below, write a definition for each vocabulary word.

8. vitamins

9. minerals

Answer the following on the lines provided.

10. Why are vitamins and minerals important to your health? Give an example of what could happen if you did not have enough of these nutrients in your diet.

11. How is water used in your body? How is water lost each day from your body?

12. How is calcium used in your body? What are three foods that are good sources of calcium?

Portfolio

What Would You Say?

13. Work with a partner. Imagine that you are friends of Janice and Karen. The two girls have asked you and your partner about vitamins and minerals. They want to know if they should get the vitamins and minerals they need from pills. What would you tell the girls about good sources of vitamins and minerals? Make up a food menu that would provide them with the vitamins and minerals they need.

Lesson 3

Calories and Food Energy

You need to know how much energy is in food. Knowing this will help you plan a healthy diet.

Lesson Objectives

You will be able to

- determine the number of calories in different foods.
- judge if you are meeting your daily energy needs.

Words to Know

diet   all the foods you eat in a day
calories   units used to measure the amount of energy in foods

Carlos and Libbie were at the library. They were trying to find out how much energy is in food.

"Which food has more energy?" asked Carlos. "An apple, a slice of bread, or a carrot?"

"I'm not really sure," answered Libbie. "I think they are about the same. One of these books should be able to tell us."

"Look," said Carlos. "Here's a table of calories in food. Is that what we're looking for?"

"Finally," exclaimed Libbie. "That's exactly what we need. Now we can figure out how much energy we get from food each day."

Measuring the Energy in Food

Now that you know about the important nutrients in food, you are almost ready to plan a healthy **diet**. Your diet is all the foods you eat during the day.

You also need to know how much energy is in the food you eat. The energy in food is measured in units called **calories**.

Different foods have different amounts of energy. So, they have different numbers of calories. A carrot, for example, has about 42 calories. An apple and a slice of bread each have about 100 calories. What about some of your favorite foods? A slice of cheese pizza has about 180 calories. A piece of fried chicken has about 250 calories. The table below shows the number of calories in some different foods.

You can find out a lot more about the calories in foods. Look in books in the library. Ask the librarian in your school for help.

Food	Number of Calories
cucumber, $\frac{1}{2}$ sliced	7
cherry tomatoes, 6	26
tossed green salad, small, no dressing	50
orange, medium	68
bagel, $\frac{1}{2}$	75
milk, 1 cup skim	80
banana, small	88
egg, boiled	100
yogurt, 1 cup nonfat	110
rice, $\frac{1}{2}$ cup cooked	110
cottage cheese, $\frac{1}{2}$ cup	120
oatmeal, 1 cup cooked	146
milk, 1 cup whole	160
French fries, 20	220
cheddar cheese, 2 ounces	230
hamburger (no roll), 3 ounces	245
tuna salad sandwich	278
vanilla milk shake, 10 ounces	350
fried fish sandwich	402

You need to know one more thing when planning your daily diet. You need to know how many calories you need to eat each day.

Different people need different amounts of calories. The number of calories a person needs depends on their sex, age, body weight, and amount of physical activity. The table at left shows about how many calories a person needs each day.

Age	Sex	Number of Calories
7–10	male and female	2,400
11–14	male	2,700
	female	2,200
15–18	male	2,800
	female	2,100
19–24	male	2,800
	female	2,100
25–50	male	2,700
	female	2,000
51+	male	2,400
	female	1,800

As you can see, males and females need different amounts of calories. Males in general need more calories each day than females because they weigh more. People of different ages also need different amounts of calories. Older people usually need fewer calories because their bodies do not use the calories in food as quickly as the bodies of younger people. No matter what your age, if you are very active, you need more calories than a person who is not active.

Carbohydrates, proteins, and fats all provide calories your body can use. But fats provide more calories than equal weights of proteins or carbohydrates. One gram of fat provides 9 calories, while one gram of protein or carbohydrate provides only 4 calories.

It is important to know how much energy is in food and how much energy you need each day. That way you can make sure you get all the energy your body needs.

Nutrition in Action

Fill in the blanks in the following statements with the correct word or phrase.

1. One gram of _____ provides 9 calories.

2. One cup of skim milk has _____ calories than 2 ounces of cheddar cheese.

3. A teenage boy usually needs _____ calories each day than a teenage girl.

4. A female between age 19 and 24 needs about _____ calories each day.

5. I need to eat about _____ calories each day.

Lesson Review

Vocabulary Review

On the lines below, write a sentence using each vocabulary word correctly.

6. calories _____

_____

7. diet _____

_____

Write **T** on the line in front of each true statement. Write **F** on the line in front of each false statement. Change each false statement to a true statement.

____ 8. Every person needs the same number of calories each day.

_____

____ 9. All kinds of food provide the same number of calories.

_____

____ 10. A cup of skim milk and a banana have about the same number of calories.

_____

____ 11. A tuna salad sandwich has more calories than a fried fish sandwich.

_____

Answer the following questions on the lines provided.

12. What two things do you need to know to plan your daily diet?

_____

_____

13. Why is it important to know the amount of energy in food and the amount of energy you need each day?

_____

_____

Portfolio

14. Find books in the library that list the number of calories in different foods. Keep track of all the foods you eat in one day. Find the number of calories for each food you ate. Find the total number of calories you ate in one day. How does the number of calories you ate compare to the number of calories you need? Keep a record of all the food and the number of calories you eat in one week.

UNIT REVIEW

Take Another Look

1. What kinds of nutrients are shown in the photo above? How are these nutrients used by your body?

   _____

   _____

   _____

2. Fill in the chart below. Write as much information as you can.

Type of Nutrient	What It Does for the Body	Food Sources
Vitamins		
Minerals		
Water		

18 Unit Review

Reviewing What You Know

Write the number of the description in Column 1 in front of the nutrient it goes with in Column 2.

Column 1	Column 2
3. help the body build and repair cells and tissues	___ a. carbohydrates
4. help the body use other nutrients	___ b. fats
5. helps break down food and carry away wastes	___ c. minerals
6. supply the body with concentrated energy	___ d. proteins
7. help the body grow and function normally	___ e. vitamins
8. the body's main source of energy	___ f. water

Answer the following questions on the lines provided.

9. Explain the difference between carbohydrates and fats. How does your body use each of these nutrients?

   _____

   _____

10. Why is it important to know how many calories are in food and how many calories you need each day?

    _____

    _____

11. How does your body use water?

    _____

    _____

Cooperative Learning

12. Work with one or two partners. Prepare a list of questions to ask the person in your school who plans or prepares school lunches. You might want to ask questions such as the following: How many calories are in each lunch? How do you select the foods to serve at lunch? Are the lunches nutritious? Make an appointment to speak with the person. Ask him or her these and other questions. Also pick a school lunch for one day. Identify the nutrients in the lunch for that day. Prepare an oral report and share your findings with your class.

13. Work with a partner. Both of you keep track of all the foods you eat for 2 days. After 2 days, analyze what you have eaten. Use information in Lessons 1–3. If you need additional information, use books in the library. What nutrients were in the foods you ate? What nutrients were missing? What vitamins and minerals were in the foods you ate? What vitamins and minerals were missing? Did you drink enough water? How many calories were in the foods you ate? Did you eat too many or too few calories? How could you improve your daily diet? Summarize your results on a poster.

UNIT TWO • MANAGING YOUR MEALS

Lesson 4

Choosing a Balanced Diet

Be sure to get fruits and vegetables when you shop for food. You should eat at least five servings of fruits and vegetables each day.

Lesson Objectives

You will be able to

- identify the six groups of food in the food group pyramid.
- choose categories of foods to ensure a balanced diet.

Words to Know

balanced diet	diet that contains a variety of foods and provides all the nutrients your body needs to stay healthy and work properly
food group pyramid	diagram that shows the six groups of foods in a balanced diet and the number of daily servings for each group
saturated fats	fats from animals; usually solids at room temperature
serving	an amount of food
unsaturated fats	fats from plants; usually liquids at room temperature

Jacob and his father were at the grocery store. They were shopping for Jacob's birthday dinner.

"Why are we having so many vegetables?" Jacob asked his father. "I thought it was my birthday and I could eat what I wanted?"

"You'll get a cake and that will be enough sugar for one day," his father answered. "Besides, I'm making baked sweet potatoes the way your grandma makes them."

"I forgot about those," said Jacob. "They're so good I forget they are vegetables. Put a few more in the basket."

"OK," said Jacob's father. "Let's also get some nice fresh green beans."

Choosing Foods for a Balanced Diet

Jacob and his father are trying to choose healthy foods for Jacob's birthday dinner. You already know that food contains many different nutrients. Your body needs six basic nutrients each day. To get all these nutrients, you need to eat a **balanced diet**. A balanced diet provides all the nutrients your body needs to stay healthy and work properly. But how do you know if you are eating a balanced diet? It takes a little work, but you can do it if you try.

One way to tell if you're eating a balanced diet is to follow the **food group pyramid**. Look at the pyramid shown on page 22. This pyramid is printed on many food packages. The food group pyramid shows the six food groups that provide important nutrients. It also tells you how much food from each group you should eat each day. You should eat more of the foods at the bottom of the pyramid than the foods at the top. Let's take a closer look at the food group pyramid.

The bottom row of the pyramid is the biggest. It contains the most important foods in a balanced diet. The bottom row of the pyramid is the Bread, Cereal, Rice, and Pasta Group. These foods contain complex carbohydrates, which are the body's main source of energy. These foods also provide other important nutrients such as vitamins and minerals. You should eat 6 to 11 servings from this group each day.

What is a **serving**? A serving is simply an amount of food. The size of a serving is different for each food in each food group. A serving in the Bread, Cereal, Rice, and Pasta Group, for example, equals 1 slice of bread or 1 cup of cooked rice. And $\frac{1}{2}$ to $\frac{3}{4}$ of a cup of cereal also equals 1 serving. You will learn more about serving size when you study food labels in Lesson 5.

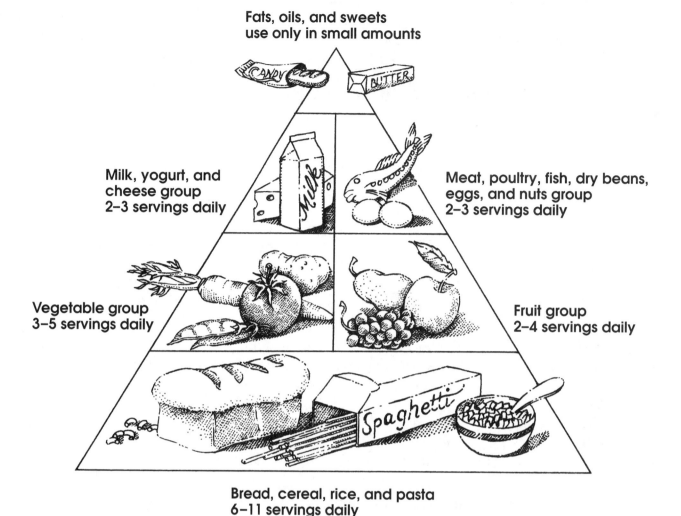

The food group pyramid shows the six food groups.

Two food groups are on the second level of the pyramid. On the left side is the Vegetable Group. This group contains all kinds of vegetables. Try to eat dark green vegetables such as broccoli, leafy vegetables such as spinach and lettuce, and yellow and orange vegetables such as carrots and squash. These vegetables have the most vitamins and minerals. You should eat 3 to 5 servings from this group each day. One serving equals one potato, one ear of corn, one tomato, or one cup of beans.

The other food group on the second level is the Fruit Group. It contains all kinds of fruits, including apples, oranges, melons, berries, and bananas. You should eat 2 to 4 servings from this group each day. One serving equals one apple, one banana, one orange, one pear, or one small bunch of grapes.

Many supermarkets and grocery stores try to help you eat better. They have "5 Each Day" signs in the fruit and vegetable section. These signs remind you that you should eat up to 5 servings each day from the Vegetable and Fruit groups. Look for these signs when you go shopping at a supermarket or grocery.

The third level of the pyramid also has two groups. As you can see, these groups are smaller in size. You should eat less from these two groups. The group on the left is the Milk, Yogurt, and Cheese Group. Foods in this group contain protein, vitamins, minerals, and some fats. You should eat 2 to 3 servings from this group each day. If possible, the foods you choose should have little or no fat such as skim milk and nonfat yogurt.

The other food group on the third level is the major source of protein in a balanced diet. It is the Meat, Poultry, Fish, Dry Beans, Eggs, and Nuts Group. That's a long name for this important group. You should eat 2 to 3 servings from this group each day. What is a serving in this group? One egg is a serving. About 3 ounces of lean meat, poultry, or fish is also a serving.

The top of the pyramid is the Fats, Oil, and Sweets Group. Foods from this group should be eaten only in small amounts. Try to avoid unnecessary fats, oils, and sweets in your daily diet.

Nutrition in Action

1. Explain what the food group pyramid is and how it is used.

   _____

   _____

   _____

2. Name the six food groups in the food group pyramid and list how many servings from each group you should eat each day.

   _____

   _____

   _____

   _____

   _____

Another way to help you plan your diet

The pie chart below shows how much food energy should come from carbohydrates, proteins, and fats. It can help you plan how much of these nutrients you should eat.

About 60 percent of the calories in your food should come from carbohydrates. Most of these should be complex carbohydrates. Suppose that your diet is 2,100 calories per day. You should eat about 315 grams of carbohydrates. This provides 1,260 calories.

No more than 30 percent of your calories should come from fats. If you take in 2,100 calories per day, 70 grams of fat is enough to provide 630 calories. This is 30 percent of your calories. There are two kinds of fats. **Saturated fats**, such as butter and cream, are fats from animals. No more than 10 percent of the calories in your food should come from saturated fats. **Unsaturated fats** come from plants. No more than 20 percent of the calories in your food should come from these fats.

The rest of the calories in your diet should come from protein. On 2,100 calories per day, about 210 calories should come from protein. About 53 grams of protein is enough.

CALORIES FROM NUTRIENTS

- Carbohydrates: 60%
- Proteins: 10%
- saturated fats: 10%
- unsaturated fats: 20%

Learn to balance the amounts of carbohydrates, proteins, and fats in your diet.

Think About It

3. Why is it important to eat a balanced diet?

_____

_____

4. What tools can you use to plan a balanced diet?

_____

_____

5. Why should most of the food you eat each day be complex carbohydrates?

_____

_____

Lesson Review

Vocabulary Review

Use words from the list below to fill in the blanks in the sentences.

balanced diet   food group pyramid   serving   saturated fat   unsaturated fat

6. There are two kinds of fats. _____ fats come from animals. _____ fats come from plants.

7. The amount of any kind of food you should eat is a(n) _____ .

8. It is important to eat a(n) _____ . One way to get all the nutrients you need is to follow the _____ .

Write the number of each food group listed in Column 1 on the line in front of the number of servings in Column 2. You can use some of the numbers of servings more than once.

Column 1
9. fats, oils, and sweets
10. vegetable group
11. milk, yogurt, and cheese group
12. bread, cereal, rice, and pasta group
13. fruit group
14. meat, poultry, fish, dry beans, eggs, and nuts group

Column 2
____ a. 2–3
____ b. 2–4
____ c. 3–5
____ d. only small amounts
____ e. 6–11

Portfolio

What Would You Do?

15. Keep a record of all the food you eat during one day. Compare the foods to the food group pyramid. Did you eat the correct number of servings from each group? What groups were you missing? How would you have to change your diet to match the number of servings from each group in the pyramid?

Lesson 5

Choosing the Right Food

Look for Nutrition Facts labels on packages of food. The information in these labels will help you choose foods with the nutrients you need.

Lesson Objectives

You will be able to

- identify the kinds of information on a food label.
- use nutritional information on food labels to select a balanced diet.

Words to Know

cholesterol	a kind of saturated fat found in foods such as meats and butter
fiber	the undigested parts of plants; keeps digestive system clean and working properly
gram	a measurement of weight
sodium	a mineral that makes up part of table salt

Tamara and Reggie were shopping at the market. They were looking for healthy food to make for dinner.

"Look at this label," said Tamara. "It sure has a lot of information on it."

"I'll say," commented Reggie. "Do you have any idea what it all means?"

"I'm not sure," said Tamara. "But, it looks like all the information is important. Look, here's information about fat."

"And here it tells you how much carbohydrate you need to eat in a day," pointed out Reggie.

"I wish we knew more about these labels," said Tamara. "Who do you think can help us?"

Nutrition Facts

All packaged food has a special label called "Nutrition Facts." Some markets also provide this information for fresh foods such as fruits and vegetables.

The next time you go to the grocery store, look for the Nutrition Facts label on packages. It looks like the one at right. The information on the label will help you choose foods with the nutrients you need. Let's take a closer look at the label from top to bottom. Look for each part of the label as it is described.

At the top of the label is the serving size for the food. The rest of the numbers on the label are based on one serving. Right below the serving size is the number of servings of food in the package.

Below the serving size information is the number of calories per serving. The label also lists the number of calories from fat in one serving.

The next section of the label shows how much of the daily amounts of fats, carbohydrates, and proteins are in one serving of the food. The first item is Total Fat. The amount is given in **grams**. A gram is a measurement of weight. (A large paper clip weighs about 1 gram.) Right below Total Fat are the amounts of saturated and unsaturated fats.

The next item is **cholesterol**. Cholesterol is a kind of saturated fat found in foods such as meats and butter. Too much cholesterol can cause diseases of the heart and blood vessels in some people. You should try to eat only a small amount of cholesterol each day.

A Nutrition Facts label

The next item of the label is **sodium**. Sodium is a mineral, and is also part of table salt. Too much sodium in your diet is not good for you. Too much salt can cause high blood pressure and other kinds of disease.

The next part of the label is Total Carbohydrates. You already know that you should eat 6 to 11 servings of carbohydrates each day. Below Total Carbohydrates is the amount of **fiber** and sugar in each serving. Fiber is the undigested parts of plants. It helps keep your digestive system clean and working properly.

The last item listed is protein. Remember, only about 12 percent of your daily diet should be protein.

The second half of the label lists the percentages of important vitamins and minerals in each serving of the food in the package. This part of the label also tells you how many grams of fat, cholesterol, sodium, and carbohydrates you need to eat each day.

The last part of the label tells you that one gram of fat provides 9 calories, one gram of carbohydrate provides 4 calories, and one gram of protein provides 4 calories.

Nutrition in Action

Answer the following questions using information from the Nutrition Facts label shown on page 27.

1. What is the serving size for this food? How many servings are in this package?

   _____

   _____

2. How much sodium is in one serving of the food?

   _____

   _____

3. How much more total fat could you eat in one day if you had one serving of this food?

   _____

   _____

Lesson Review

Vocabulary Review

Use words from the list below to fill in the blanks in the paragraph.

cholesterol   fiber   grams   sodium   Nutrition Facts   serving

The label on packaged food is called (**4.**) _____. It shows how much of important nutrients are in one (**5.**) _____. The amounts are given in (**6.**) _____. One nutrient listed is (**7.**) _____, a kind of saturated fat. Another important nutrient is (**8.**) _____, the undigested parts of plants. The label also lists the amount of the mineral (**9.**) _____, which is a part of table salt.

Write answers to the following questions on the lines provided.

10. How can you use the information in Nutrition Facts labels to help plan a balanced diet?

_____

_____

_____

11. Look at the Nutrition Facts label on page 27. If you ate one serving of the food, how many more grams of fiber should you eat during the day?

_____

12. If you ate one serving of the food, how many more grams of cholesterol could you eat during the day?

_____

Portfolio

What Would You Do?

13. Work with a partner. Go to the grocery store or look at packages of food in your kitchen at home. Study the Nutrition Facts labels on packages of different kinds of foods. Try to plan a meal—lunch, for example—that supplies some of all the basic nutrients you need. Use the information in the labels. What foods did you select? In your journal, write why you think studying Nutrition Facts labels is important.

Lesson 6
Meal Planning

A balanced meal has a variety of foods. It's important to plan balanced meals.

Lesson Objectives

You will be able to

- plan a balanced meal and a balanced diet.
- avoid excess fat, salt, and cholesterol.
- avoid eating junk food.

Words to Know

balanced meal	meal that contains a variety of foods and nutrients; part of a balanced diet
fast food	foods that can be served quickly from restaurants; many are fried
junk food	common name for foods that contain lots of calories but not many nutrients
snack	anything you eat or drink between meals

May-lei, her brother Kim, and her parents were having dinner. It was unusual for them all to be home at the same time.

"This is almost like a celebration," said May-lei. "I can't believe we're all eating dinner together."

"How long has it been?" asked Kim.

"Not that long," said May-lei's mother. "It just seems that way. It was fun to make us all a special dinner tonight."

"The food is all delicious," commented May-lei's father. "This fish is my favorite."

"Mine, too," said May-lei. "And everything is so fresh."

"And good for you," said Kim as he laughed. "I'm studying health and nutrition at school right now."

"Me too," said May-lei. "And this is a balanced meal if I ever saw one."

Planning a Balanced Meal

The meal May-lei's mother prepared was a **balanced meal**. A balanced meal has a variety of foods that provide important nutrients. The meal had protein from fish, vitamins and minerals from the vegetables, and carbohydrates from the rice. The meal was low in fat and high in fiber.

You know it is important to eat a balanced diet. Learning to eat a balanced diet takes time and practice. You need to remember everything you have learned about nutrients. You also need to remember the food groups and how many servings to eat from each group each day. That's a lot to remember and do. So, the best thing to do is start slowly. Begin by planning and eating one balanced meal at a time. Before long, you'll be eating three balanced meals each day.

Sometimes you get hungry between meals. Most people do. When planning your diet, don't forget about snacks. A **snack** is anything you eat or drink between meals. You should make sure your snacks are part of a nutritious and balanced diet. What do you normally have for a snack? Potato chips? A soft drink? A candy bar or cookies? Try to replace these snacks with healthy snacks such as fruit, yogurt, juice, chips that are baked (not fried), and air-popped popcorn. These snacks have vitamins, minerals, and other nutrients. They taste good, too! Remember, everything you eat and drink during the day is part of your diet.

Think About It

1. What is a balanced meal? How does eating a balanced meal help you have a balanced diet?

2. Why is it important to eat healthy snacks?

Five diet goals

Here are five simple guidelines to keep in mind when planning a balanced meal and diet.

1. Eat a variety of foods. Eat foods from all the different food groups. These foods provide the different nutrients you need. Planning a balanced diet involves trying new and different kinds of foods. Many people enjoy finding about new ways to eat food. They also like to learn about the foods people eat in other countries. Try something different, maybe a new vegetable or fruit. You might actually like what you try.

2. Avoid eating too much fat, sodium, and cholesterol. These nutrients are found in many **junk foods** and **fast foods**. Junk foods, such as potato chips, cookies, and soft drinks, have lots of calories but not many nutrients. Fast foods, including hamburgers, tacos, and french fries, are the foods you pick up from a restaurant and eat quickly. Many of these foods are fried and have lots of fat. Many people enjoy the taste of these foods. You don't need to avoid them completely. But, try to limit how much of them you eat. Instead, eat fresh foods if possible.

3. Don't eat too much sugar. You have already learned that sugar is a simple carbohydrate. Foods that contain mostly sugar have lots of calories but not many other nutrients. Too much sugar in the diet can make you overweight. It can also cause tooth decay. To reduce the amount of sugar in your diet, avoid soft drinks, candy, donuts, and cookies. Eat more fresh fruit and drink juice and water. Read food labels to find out how much sugar is in food.

Which foods should you choose for a healthy snack?

4. Eat enough carbohydrates and fiber. Remember to eat 6 to 11 servings of carbohydrates each day to have the energy your body needs.

5. Eat only the amount of calories you need. In Lesson 3, you learned that different people need different amounts of calories each day. Look back at the chart on page 16 to find out about how many calories are right for you. You might also talk to your doctor or a school nurse about the number of calories you should eat, based on your age, weight, and activity. Try not to eat more than that amount. If you eat too many calories, you may gain weight. Eating the right number of calories each day provides the right amount of energy your body needs to stay active.

Three balanced meals a day

Many people say that breakfast is the most important meal of the day. Think about why this is true. When you get up in the morning, it has been a long time since your last meal. It may be 12 hours or more since you ate dinner. A nutritious breakfast provides energy for the activities you will do the rest of the day. Without a nutritious breakfast, you feel tired and do not have enough energy to stay active. Sometimes you don't think clearly. A nutritious breakfast will help you feel alert and ready for the day.

What makes up a good breakfast? Like any meal, you should eat a variety of foods from the different groups. Nutritious breakfast foods include whole-grain cereal, fruit and fruit juice, low fat or nonfat milk and/or yogurt, toast with jam, and maybe an egg two or three times each week. Avoid eating meats such as sausage and bacon. Also skip the donuts and sweet rolls. A healthy breakfast should provide about one-quarter of your daily nutrients and calories.

Juice, fruit, whole-grain cereal, and toast are good foods to include in your breakfast.

If you eat a nutritious breakfast, you might not be hungry until lunch. You can eat many different kinds of food for lunch. What are some of your favorite foods for lunch? Are they part of a balanced diet? Try to eat one or two servings of foods that contain protein at lunch. And don't forget to eat servings of fruits, vegetables, and bread or pasta. Healthy choices for lunch include salads with low-fat dressing, a lean hamburger with no cheese, or a turkey or chicken sandwich on whole-grain bread. A peanut butter sandwich or a bowl of spaghetti are also good lunch choices. Soup is also a good lunch food. Avoid eating fatty foods such as French fries, potato chips, or other fried foods.

You also have many choices of food to eat at dinner. Many people plan their dinners around a large piece of meat. Try to reduce the amount of meat you eat at dinner. Instead, increase the number of servings of bread, rice, pasta, fruits, and vegetables you eat. Throughout the day, avoid eating foods that contain too much fat and sugar. And remember to drink plenty of water.

Nutrition in Action

3. Why is it important to avoid eating too much fast food and junk food?

___

___

4. Why is it important to eat a variety of foods each day?

___

___

5. Why is breakfast the most important meal of the day?

___

___

Lesson Review

Vocabulary Review

On the lines provided, write a sentence that explains the connection between each pair of words.

6. balanced meal, snack

   _____

   _____

7. junk food, fast food

   _____

   _____

Fill in the blank in the following sentences with the correct word or phrase.

8. Anything you eat or drink between meals is a _____.

9. Avoid eating too much _____, _____, and _____.

10. Eating a _____ of foods gives you the nutrients you need each day.

Answer the following questions on the lines provided.

11. What are five diet goals to keep in mind when planning a balanced diet?

    _____

    _____

    _____

    _____

12. Why is it important to be sure your snacks are nutritious?

    _____

    _____

Portfolio

What Would You Do?

   Work with a partner. Use all the information you have learned so far about nutrition. Use cookbooks from home and books from the library. Plan three balanced meals and two snacks for a family with two adults and two teenagers. What will the family eat for breakfast, lunch, and dinner? Is everyone in the family eating all the nutrients they need? Don't forget about things to drink. Make a poster showing the balanced meals for the day. Share your findings with your class.

UNIT REVIEW

Take Another Look

1. What food groups are included in the meal shown above? How many servings from each group are included in this meal?

   _____
   _____
   _____

2. Is this meal a balanced meal? Explain your answer.

   _____
   _____

3. Imagine you had this meal for dinner. What would you eat for breakfast, lunch, and a snack to have a balanced diet for the day? How many servings from each of the six food groups would you have to eat?

   _____
   _____
   _____
   _____
   _____

Reviewing What You Know

Complete the following sentences.

4. About 58 percent of the food you eat each day should come from _____ .

5. _____ fats, such as butter and cream, come from animals. Fats that come from plants are called _____ .

36 Unit Review

6. All packaged foods have a special label called _____.

7. The undigested parts of plants that keep your digestive system clean are called _____.

Answer the following on the lines provided.

8. From bottom to top, list the six food groups in the food group pyramid and how many servings from each group you should eat each day.

   _____
   _____
   _____
   _____
   _____

9. Why is it important to eat a balanced diet each day?

   _____
   _____

10. How can you use the information in the food group pyramid to help plan a balanced diet?

    _____
    _____

11. How can you use the information on food labels to help plan a balanced diet?

    _____
    _____

Cooperative Learning

12. Work with one or two partners. Use books in the library to find out what kinds of food people in different countries eat. Look up countries such as China, Japan, Australia, Mexico, Italy, France, Turkey, and Germany. Find out what people in the country would eat in a typical day. What is their main source of protein? From what foods do they get carbohydrates? How are their foods different from the foods you eat? How are they similar? If possible, try to cook some of the foods from the country you selected.

13. Work with a partner. Create a table with the names of the six food groups across the top. On the left side of the table, write the days of the week: Monday through Friday. Each day for one week, record the foods served for lunch in the school cafeteria. Also record the number of servings from each food group for each day. At the end of the week, study your results. Were the lunches served at your school balanced meals? Were all the food groups included in each lunch? Which food groups were missing? What kinds of foods could be added to make the lunches more balanced? Summarize your findings. Write a letter to the person in charge of planning school lunches. Tell the person what you found out from your study.

UNIT THREE • MANAGING YOUR WEIGHT
Lesson 7
Weight and Nutrition

A person's weight depends on age, sex, height, and build.

Lesson Objectives

You will be able to

- understand the importance of a healthy weight range for each individual.
- understand the possible dangers of falling outside that range.

Words to Know

anorexia nervosa	condition in which a person starves him or herself
bulimia	condition that involves eating large amounts of food followed by starvation or vomiting
build	the shape and size of your body, including the amount of muscle, bone, and fat that make it up
ideal weight	the weight recommended for your age, sex, height, and build
obese	having excess body fat
obesity	being more than 20 percent over the body weight shown on height and weight tables
overweight	being more than 10 percent over the body weight shown on height and weight tables

Rachel and Celia were in Rachel's bathroom.

"I can't believe I weigh 128 pounds," cried Rachel. "I've never weighed this much in my life! How much do you weigh, Celia?"

Celia stood on the scale. "Look, I only weigh 107 pounds," said Celia.

"I can't believe I weigh so much," said Rachel. "What can I do to lose some weight?"

"How do you know you weigh too much?" asked Celia. "You don't look fat at all."

"But I weigh more than 20 pounds more than you do," said Rachel.

"I know," replied Celia. "But, you're also about 5 inches taller than I am. That might make a difference, you know."

"Maybe you're right," said Rachel. "Even your feet are smaller than mine."

Both girls laughed.

How Much Should You Weigh?

Rachel and Celia just discovered some important facts. Not all people weigh the same. Not all people are the same height and **build**. A person's build is the shape and size of his or her body, including the amount of muscle, bone, and fat. Because people come in different sizes and shapes, they also have different weights. A person's weight depends on his or her age, sex, height, and build. But, how do you know if you weigh what you are supposed to?

The table on page 40 lists ranges for the **ideal weights** of boys and girls your age. Your ideal weight is how much you should weigh based on your height, build, age, and sex. Notice that a range of weights are shown on the table. For example, the ideal weight for a male 6 feet tall with a medium build is between 157 and 170 pounds.

Nutrition in Action

1. What is your ideal body weight based on?

2. Based on the table that follows, what is your ideal weight range?

Should you weigh what the table says?

Tables of ideal weights are made by experts who look at the heights and weights of thousands of people. But the weight range on the table might not be right for you. The table does not tell you how much of your weight is bone, how much is muscle, and how much is fat. According to the table, a very muscular person might be **overweight**, or more than 10 percent over the weight shown on the table. But remember, you can be overweight on the table if you have large bones and muscles because bones and muscles weigh more than fat. On the other hand, a person who gets little exercise might not be overweight according to the table. But that person might have too much body fat. People who are 20 percent heavier then their weight shown on the table with excess body fat are **obese**.

Ideal Body Weights (without clothes)

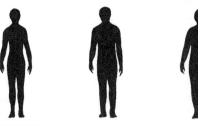

Height (without shoes)	Light Build	Medium Build	Heavy Build
MALES			
5 ft. 4 in.	132–138	135–145	142–156
5 ft. 5 in.	134–140	137–148	144–160
5 ft. 6 in.	136–142	139–151	146–164
5 ft. 7 in.	138–145	142–154	149–168
5 ft. 8 in.	140–148	145–157	152–172
5 ft. 9 in.	142–151	148–160	155–176
5 ft. 10 in.	144–154	151–163	158–180
5 ft. 11 in.	146–157	154–166	161–184
6 ft.	149–160	157–170	164–188
6 ft. 1 in.	152–164	160–174	168–192
6 ft. 2 in.	155–168	164–178	172–197
6 ft. 3 in.	158–172	167–182	176–202
6 ft. 4 in.	162–176	171–187	181–207
FEMALES			
4 ft. 10 in.	102–111	109–121	118–131
4 ft. 11 in.	103–113	111–123	120–134
5 ft.	104–115	113–126	122–137
5 ft. 1 in.	106–118	115–129	128–143
5 ft. 2 in.	108–121	118–132	128–143
5 ft. 3 in.	111–124	121–135	131–147
5 ft. 4 in.	114–127	124–138	134–151
5 ft. 5 in.	117–130	127–141	137–156
5 ft. 6 in.	120–133	130–144	140–159
5 ft. 7 in.	123–136	133–147	143–163
5 ft. 8 in.	126–139	136–150	146–167
5 ft. 9 in.	129–142	139–153	149–170
5 ft. 10 in.	132–145	142–156	152–173
5 ft. 11 in.	135–148	145–159	155–176
6 ft.	138–151	148–162	158–179

How do you know if you have too much body fat?

Some parts of your body are more likely to store fat than others. Two places where it is easy to spot extra fat are just above your waist and at the top of your arm, just below your armpit. You can try this test on those two areas of your body. With your thumb and forefinger, try to pinch as much flesh as you can. If you can pinch a thickness of more than $\frac{3}{4}$ inch, you may have excess body fat.

What happens if you have too much body fat?

Being obese can cause many serious health problems. People who are obese often don't live as long as people who are their ideal weight. The chance of dying before age 60 is much greater for people who are obese. Excess fat can build up in blood vessels and prevent blood from flowing properly. Blocked blood vessels can cause heart attacks or other problems. Since the heart must work harder in obese people, they usually tire more easily. Obese people are also more likely to develop high blood pressure, certain cancers, and other diseases. You can reduce your chances of having these health problems as an adult if you reduce your body fat as a teenager.

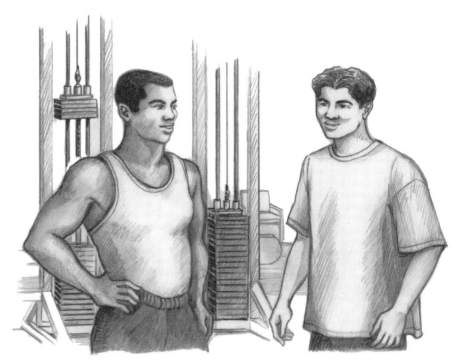

These teens are neither overweight nor underweight.

What happens if you do not weigh enough?

If you open up any magazine or watch TV, you see attractive people who are very thin—some are even skinny. Unfortunately, many people—especially teenage girls—think they have to be thin to be liked.

Many teenagers—both boys and girls—try to weigh less than they should. They think they look better if they weigh less. Sometimes, these people just don't eat enough food. They skip meals or eat only tiny amounts of food. Some of these people have conditions that rob the body of the nutrients it needs to grow and stay healthy.

One of these conditions is called **anorexia nervosa**. This is an emotional disorder in which a person starves himself or herself by not eating. This condition is much more common in teenage girls than in teenage boys. These people eat smaller and smaller amounts of food until they eat almost nothing at all.

Another serious condition is **bulimia**. People with bulimia eat large amounts of food and then cause themselves to vomit. Some people with anorexia or bulimia also exercise too often and too hard.

Lesson 7   41

Both anorexia nervosa and bulimia cause serious health problems. The body does not get the nutrients it needs for energy or to work well. Sometimes the heart does not beat regularly. The body is too weak to fight off infections. Also, forced vomiting causes tooth and throat damage. People with anorexia or bulimia usually need medical help to overcome these problems.

Think About It

3. What kinds of health problems might a man with a light build who is 5'9" and weighs 180 pounds have? Why?

   _____

   _____

   _____

4. What kinds of health problems might a woman with a medium build who is 5'3" and weighs 105 pounds have? Why?

   _____

   _____

   _____

Maintaining your ideal weight

In Lesson 3 you learned that the amount of energy in food is measured as calories. You also learned that different foods have different amounts of calories. To maintain your ideal weight, you need to understand how the number of calories you eat and the number of calories your body uses are related. Simply put, if you eat the same number of calories as your body uses each day, your weight stays the same. If you eat more calories than your body uses, you gain weight. If your body uses more calories than you eat, you lose weight. You will learn more about calories and weight in Lesson 8.

It is important to maintain your ideal body weight. When you maintain your ideal weight, you better your chances to live a longer and healthier life. Maintaining your ideal weight keeps you looking and feeling your best and helps you avoid the health problems you read about. In addition, you will be less likely to have problems with your joints, muscles, and back if you maintain your ideal weight.

Lesson Review

Answer the following on the lines provided.

5. Explain why all people do not have the same ideal weight.

_____

_____

6. How are the eating disorders anorexia nervosa and bulimia similar? How are they different?

_____

_____

7. How can you find out if you have too much body fat?

_____

_____

Vocabulary Review

Use words from the list below to fill in the blanks in the paragraph.

anorexia nervosa   bulimia   build   ideal weight   obese   overweight

Your (**8.**) _____ is the number of pounds you should weigh. This number is based on your height, (**9.**) _____ , age, and sex. Sometimes people weigh more than they should. People who weigh more than 10 percent over the weight shown on the table are (**10.**) _____ . People with too much body fat are (**11.**) _____ . Sometimes people do not weigh enough. A condition that causes starvation is called (**12.**) _____ . Another condition in which a person eats a lot of food and then vomits is called (**13.**) _____ .

Portfolio

What Would You Say?

14. Suppose your friend is 5'10" tall, has a medium build, and weighs 190 pounds. He is a wrestler who does weight training. He is concerned because his weight is more than the ideal weight shown on the table. What would you tell him? How could he decide if he needs to lose weight?

Lesson 8
Losing Weight Safely

Exercise is part of a safe plan to lose extra weight.

Lesson Objectives

You will be able to

- identify safe methods for losing weight.
- make a plan to reduce weight, if necessary.

Words to Know

calorie intake	the total number of calories you eat and drink in one day
calorie output	the number of calories you burn during one day
diet	a program to lose weight

Hunter had not been feeling very well. He was tired all the time and just didn't have any energy. He went to see his school nurse.

"There's nothing wrong with you, Hunter, that a little exercise won't cure," said the nurse.

"What do you mean?" asked Hunter. "I'm too tired to exercise. That's why I came to see you."

"What I'm trying to say," the nurse said, "is that you are overweight. You need to lose about 10 pounds. Then you'll feel better."

"How can I do that?" asked Hunter.

"Easy," said the nurse. "You need to get more exercise and eat less food but more nutritious food each day. Let's get you started on losing weight."

"OK," said Hunter. "I'll get started today."

The Right Way to Lose Weight

Hunter did a wise thing. He went to see a health care professional because he didn't feel well. The nurse told Hunter he needed to lose some weight. And the nurse would help him do it.

If you are overweight or obese, you will need a plan to reduce your weight. But, before you begin any kind of **diet**, a program to lose weight, see a doctor, school nurse, or someone else who has health-care training first. A health care professional can tell you if there are any things you should watch out for. You can also find out how much weight you should lose and a safe way to do it. Here is some useful information to help you lose weight safely.

In Lesson 7, you learned the relationship between the number of calories you eat and the number of calories you use. Your **calorie intake** is the total number of calories you eat and drink in one day. Your **calorie output** is the number of calories you use in one day. If your calorie intake and calorie output are the same, your weight will not change. But, if your calorie intake is less than your calorie output, you can lose weight.

How many calories do you need?

Your body uses energy every minute of the day—even when you are sleeping. So, you need to eat a certain amount of food to have enough energy to do everything your body needs to do. You also have to have enough energy to stay healthy. This amount depends on your age and sex, your body size and weight, and how active you are. Young men between the ages 15 and 20 have high energy needs. They generally need about 2,800 calories each day to stay healthy. Young women the same age need about 2,100 calories each day.

The number of calories you need changes as you grow older. Because your body has finished growing, you need fewer calories when you are an adult than you do now. The table on page 16 in Lesson 3 shows the number of calories people of different ages need each day.

Think About It

1. Why should you see a doctor or other health care professional before you begin to lose weight?

   _____
   _____
   _____

2. What is the difference between your calorie intake and calorie output?

   _____
   _____
   _____

3. Why do different people need different amounts of energy in a day?

   _____
   _____
   _____

How many calories do you use?

You know that your body burns calories all the time. You burn calories every time your body does something: each time you breathe, each time your heart beats, even each time you blink your eyes. Different activities, however, use different amounts of energy. Running a race burns many more calories than sitting and reading a book. The table on page 47 shows how many calories you burn in an hour doing some common activities.

Exercising to lose weight

To lose weight safely, you need to decrease the number of calories you eat each day. Eating 500 fewer calories each day for one week (500 × 7) equals 3,500 fewer calories. Each time you eat 3,500 fewer calories, you lose 1 pound of body weight.

Calories Burned by Different Activities

Activity	Calories burned in an hour
sleeping	100
reading	150
eating	170
walking slowly (1.5 miles)	210
gardening	220
golf	230
bicycling (6 miles)	250
walking quickly (3 miles)	350
roller skating	400
ice skating	400
tennis	420
aerobic dancing	500
cross-country skiing (8.5 miles)	600
swimming	600
running (8–10 miles)	900

To reduce the number of calories you take in, eat less meat, fats, oils, sugar, and dairy products (unless they're nonfat). Avoid eating fried foods. Eat more fruits, vegetables, and grains.

Another way to lose weight is to increase your daily activity. For each additional 3,500 calories you burn in one week, you will lose 1 pound.

What will happen if you eat fewer calories *and* increase your activity? You will lose weight ever faster.

Nutrition in Action

4. What changes in your diet can you make to help you lose weight?

_____

_____

5. What exercises could you do in one week to lose one pound?

_____

_____

6. Pick one activity from the list above. How long would you need to do that activity to lose one pound?

_____

_____

Your Weight Control Plan

Once you have decided to lose weight, you should have a plan. Here are some ideas that will help you lose weight.

1. Make a list of the reasons you want to lose weight. Decide on your goals and the amount of weight you want to lose. You should not try to lose more than one or two pounds each week. You might want to join a group of other people who are trying to lose weight.

2. Keep a journal. Write down your goals. What other goals might you have besides losing weight? One goal might be to improve your eating habits. Another goal might be to avoid eating between meals. Another goal might be to get more exercise.

3. Each day for two or three weeks, write in your journal what you eat each day. Be sure to include everything you eat and drink. Pay attention to the sizes of the servings you have. Figure out how many calories you eat each day. Write when and where you eat. For example, do you eat while watching TV or talking on the phone? Also write how much exercise you get each day. You might also want to write about how you feel.

4. Avoid overeating. Here are three simple things you can do to prevent overeating. Skip fattening foods such as French fries and butter. Eat low-calorie foods instead of high-calorie foods—carrots instead of potato chips, for example. Eat smaller amounts of food.

5. Keep track of your progress. Use your journal. Write how you feel when you meet your goals. Also write how you feel if you don't meet your goals. Don't be too hard on yourself if you don't meet all your goals. Be flexible, but steady, about your weight loss.

Here are some other tips to help you lose weight.

- Don't eat because you are nervous, upset, or angry.
- Eat only at a designated place—the kitchen table, for example.
- Don't eat while watching TV or reading.
- Eat slowly.
- Broil, bake, grill, or steam your food.
- When eating out, have sauces, gravies, and salad dressing served on the side or not at all.
- Eat healthy snacks.

Lesson Review

7. Circle all the ways listed below that can help you lose weight.
   a. Eat fewer calories each day than you use.
   b. Eat only vegetables and fruits.
   c. Make a commitment to lose weight.
   d. Eat more calories each day than you use.
   e. See a doctor, school nurse, or other health care professional before you begin a diet and exercise program.

Answer the following questions on the lines provided.

8. What are three tips to help you lose weight?

   _____
   _____

9. What are five things that should be part of your weight loss plan?

   _____
   _____
   _____

10. How does the relationship of your calorie intake and calorie output help you lose weight?

    _____
    _____

Vocabulary Review

11. Write a sentence that explains the difference between calorie intake and calorie output.

    _____
    _____

Portfolio

What Would You Do?

12. Imagine that you want to gain or lose weight. In your journal, describe how you would go about changing weight. Write the steps of your weight loss or weight gain plan. What kinds of food would you eat? What kinds of activities would you do? Who would you talk to about changing weight? How do you feel about losing or gaining weight?

Lesson 9
Avoiding Fad Diets

Diets that promise quick weight loss are not a safe way to lose weight.

Lesson Objectives

You will be able to

- describe the risks of unsafe diets.
- recognize if a particular diet plan does not meet basic nutritional needs.

Word to Know

fad diet   diet that promises quick weight loss

Roberto and Jason were reading magazines on the bus ride home from school.

"Look at this ad," said Roberto. "Do you think this guy really lost all that weight in just one week?"

"I doubt it," answered Jason. "Look at the pictures. I'm not sure they're both of the same guy."

"I think you're right," said Roberto. "Who do you think buys this stuff anyway?"

"People will buy anything," said Jason. "But they'll be wasting their money on this way to lose weight. Fad diets don't work. Or if you do lose weight, it doesn't stay off for very long."

"Really?," asked Roberto.

"Definitely," replied Jason, "To lose weight and keep it off, you need to eat right and exercise more."

Losing Weight the Wrong Way

Some people want to lose weight and will try just about anything. Sometimes they lose weight too quickly. Sometimes they do not eat enough food or the right kinds of food. They do not get the nutrients their bodies need to work properly and stay healthy. They do not get enough exercise to help them lose weight.

In Lesson 8, you learned a safe way to lose weight. Use more calories each day than you eat. This way to lose weight takes time. Some people do not want to take the time to lose weight safely. They want results immediately. They didn't become overweight overnight. They won't lose weight overnight either.

People who want to lose weight quickly look for diets that promise "quick weight loss." A **fad diet** is a diet that promises quick weight loss. A fad is an interest or activity, like a certain diet program, popular for only a short time. After a while, people find out the diet does not work and they try something else. In most fad diets, you lose weight by changing what you eat. In one fad diet, for example, you eat five grapefruits each day. In another fad diet you drink 12 glasses of fruit juice a day. As you can see, a fad diet is not a balanced diet.

People often lose weight on fad diets. But they are damaging their health. They do not get enough calories each day to supply the energy they need. People on fad diets often feel tired and cranky. They do not get all the nutrients they need for their bodies to work properly. As soon as they stop the diet, they often gain back the weight they lost, sometimes more. This happens because fad dieting causes the body to react as if it were being starved. When the person stops dieting and starts eating normally, the body turns the food into fat and stores it. The body does this to make sure it will not starve if the person starts dieting again.

Nutrition in Action

1. Would a diet that tells you to eat nothing but fruit for the first 10 days, then nothing but starches for the next 10 days, be a fad diet? Why or why not? Would this diet be healthy?

   _____

   _____

   _____

What is the best way to lose weight?

Fad diets are not very effective. As soon as people stop the diet, they gain back the weight they lost. They may then go on another fad diet to lose the weight they gained back. An unhealthy pattern of dieting, gaining weight, and dieting again may result.

The best way to lose weight and maintain the weight you want is to develop good eating and exercise habits. Knowing what kinds of food you need to eat to stay healthy is the first step. Eat a balanced diet of nutritious meals and snacks. Get plenty of exercise. Over time, you will lose weight. And you will gain a lifetime of healthy eating habits that will help you maintain your weight and health throughout your life.

Think About It

2. Why are fad diets not very effective?

   _____

   _____

Lesson Review

Write **T** on the line in front of each true statement. Write **F** on the line in front of each false statement. Change each false statement to a true statement.

_____ 3. Fad diets are a very effective way to lose weight.

_____

____ 4. Fad diets work by teaching a person how to select nutritious foods.

____ 5. A safe way to lose weight is to use more calories each day than you eat.

____ 6. A fad diet involves getting enough exercise each day.

Answer the following on the lines provided.

7. Describe the best way to lose weight and maintain the weight you want throughout your life.

8. Why do fad diets result in an increase of body fat and weight gain?

Vocabulary Review

On the lines provided, write a sentence using the vocabulary word.

9. fad diet

Portfolio

What Would You Do?

10. Work with a partner. Look in magazines and newspapers for advertisements for diets that promise quick weight loss. Place these ads on a poster. Answer the following questions about each ad. How much weight do they promise you will lose? How long will it take? How much does the diet cost? How does the diet work? Can you get your money back if the diet does not work?

UNIT REVIEW

Take Another Look

1. What is the boy doing in the picture above to help him lose weight safely?
   _____
   _____
   _____

2. Which activity shown above burns more calories? How does doing this activity help you to lose weight?
   _____
   _____
   _____

Reviewing What You Know

Circle the letter or letters of the answers that correctly complete the sentences.

3. Two serious conditions that cause people to lose weight are
   _____ and _____ .
   a. fad diets   **b.** anorexia nervosa   **c.** bulimia   **d.** obesity

4. If you weigh more than 10 percent over the ideal body weight on the height and weight table, you are _____ .
   a. obese   b. overweight   c. within your ideal range
5. A healthy amount of weight to lose in a week is about _____ pounds.
   a. 10   b. 5   c. 1–2

Answer the following on the lines provided.

6. Explain the difference between being overweight and being obese. Tell how each condition affects your health.

   _____
   _____
   _____

7. Compare your calorie intake and calorie output. What is the relationship between them if you want to keep your weight the same?

   _____
   _____

8. How does the relationship between your calorie intake and calorie output change if you want to lose weight?

   _____
   _____

9. Describe five things to keep in mind when making your weight-loss plan.

   _____
   _____
   _____

Cooperative Learning

10. Work with a partner. Use books in the school or public library. Find out more about anorexia nervosa and bulimia. What are the long-term effects for people who suffer from these conditions? How are they treated? What kinds of support groups are available for people who have these conditions? Summarize your findings in a written report to share with your class.

11. Work with two partners. Use books in the school or public library. Each of you should find information about healthy diets that help you lose weight. How are the diets different? How are they similar? What kinds of food do you eat in each diet? Why do you think the diets should work? Prepare posters that describe the main parts of each diet. Share the posters with your class.

GLOSSARY

amino acids (ah-MEE-noh AS-ihdz) *n.* units that combine to form proteins 6

anorexia nervosa (a-nuh-REHK-see-uh ner-VOH-suh) *n.* eating disorder in which a person starves himself or herself 41

balanced diet (BAL-uhnst DEYE-uht) *n.* diet that contains a variety of foods; provides all the nutrients your body needs 21

balanced meal (BAL-uhnst MEE-uhl) *n.* meal that contains a variety of foods and nutrients 31

build (bild) *n.* the appearance of your body due to the size of your muscles and bones 39

bulimia (buh-LEE-mee-uh) *n.* eating disorder that involves eating large amounts of food and then starvation or vomiting 41

calorie intake (KAL-uh-ree IN-tayk) *n.* the total number of calories you eat and drink in one day 45

calorie output (KAL-uh-ree OWT-put) *n.* the number of calories you burn during one day 45

calories (KAL-uh-reez) *n.* units used to measure the amount of energy in foods 15

carbohydrates (kar-bow-HEYE-drayts) *n.* nutrients that are the body's main source of energy 3

cells (selz) *n.* the smallest parts that make up a living thing 6

cholesterol (kuh-LEHS-tuh-rawl) *n.* a kind of saturated fat found in foods such as meat and butter 27

diet (DEYE-uht) *n.* all the foods you eat in a day 15; a program to lose weight 45

fad diet (fad DEYE-uht) *n.* diet that promises quick weight loss 51

fast food (fast food) *n.* foods that can be served quickly by restaurants 32

fats (fats) *n.* nutrients that supply the body with concentrated energy 3

fiber (FEYE-buhr) *n.* the undigested parts of plants; keeps digestive system clean and working properly 28

food group pyramid (food groop PIHR-uh-mid) *n.* shows the six food groups and number of servings in a balanced diet 21

gram (gram) *n.* a measure of weight 27

ideal weight (eye-DEE-uhl wayt) *n.* the weight recommended for your age, sex, height, and build 39

junk food (juhnk food) *n.* common name for foods that contain lots of calories but not many nutrients 32

minerals (MIHN-uh-ruhlz) *n.* nutrients that help the body grow and function 10

nutrients (NOO-tree-uhnts) *n.* substances in food that your body uses to grow and work properly 3

nutrition (noo-TRIH-shuhn) *n.* all the ways the body takes in and uses nutrients to stay healthy 3

obese (oh-BEES) *adj.* being 20 to 30 percent over your ideal body weight 40

overweight (oh-ver-WAYT) *adj.* being 15 percent over the range of your ideal weight   40

oxygen (AHKS-uh-juhn) *n.* a gas in the air you breathe   10

proteins (PROH-teenz) *n.* nutrients that help build and repair cells and tissues in the body   3

saturated fats (SACH-uh-rayt-uhd fats) *n.* fats from animals   24

serving (SER-vihng) *n.* an amount of food   21

snack (snak) *n.* anything you eat or drink between meals   31

sodium (SOH-dee-uhm) *n.* another name for salt   28

unsaturated fats (uhn-SACH-uh-rayt-ed fats) *n.* fats from plants   24

vitamins (VEYE-tuh-muhnz) *n.* nutrients that help the body use other nutrients   9

RESOURCES

Here are some materials you can use for more information.

Video

A Taste of Health: The Connection Between Eating and Health. Intellimation, Santa Barbara CA, 1991.

Eating Healthy for Weight Control. Turner Multimedia, 1990.

Food Labels. United Learning, Niles IL, 1994.

The Food Guide Pyramid. Cambridge Education, Charleston WV, 1993.

Books

A Guide to the Food Pyramid, by Shirleigh Moog. Crossing Press, Freedom CA, 1993.

Coping with Diet Fads, by June Kozak Kane. Rosen Publishing Group, New York, 1990.

Eat Right the E.A.S.Y. Way! by Joan Salge Blake. Prentice Hall Press, New York, 1991.

Eating Well with the Food Guide Pyramid. Meredith Books, Des Moines IA, 1996.

Everything You Need to Know About Diet Fads, by Karen Bornemann Spies. Rosen Publishing Group, New York, 1993.

How the New Food Labels Can Save Your Life, by Peg Jordan. Michael Wiese Productions, Studio City CA, 1992.

Teen Health the Natural Way, by Yaakov Berman. Pitspopany, New York, 1995.

Weight: A Teenage Concern, by Elaine Landau. Lodestar, New York, 1991.

Here are some organizations you can contact for more information.

The American Dietetic Association
216 W. Jackson Boulevard
Chicago IL 60606-6995
(312) 899-0040

The American Dietetic Association has a catalog of materials for consumers and professionals in nutrition. Ask about their consumer education materials.

Food and Nutrition Information Center
National Agricultural Library
United States Department of Agriculture
10301 Baltimore Blvd., Room 304
Beltsville MD 20705-2351
(301) 504-5719
fax (301) 504-6409
e-mail address: fnic@nalusda.gov

The Food and Nutrition Information Center (FNIC) produces materials about food and nutrition topics. Single copies are available in print or on floppy disk. You can also get FNIC publications via the World Wide Web at http://www.nalusda.gov/fnic/

National Dairy Council
10255 Higgins Road, Suite 900
Rosemont IL 60018
(847) 803-2000